BRITISH CATTLE

Val Porter

SHIRE PUBLICATIONS

Published in Great Britain in 2009 by Shire Publications
Ltd, Midland House, West Way, Botley, Oxford OX2 0PH,
United Kingdom.
443 Park Avenue South, New York, NY 10016, USA.
E-mail: shire@shirebooks.co.uk   www.shirebooks.co.uk

© 2001 Val Porter. This new edition 2009.

Every attempt has been made by the Publishers to secure
the appropriate permissions for materials reproduced in
this book. If there has been any oversight we will be happy
to rectify the situation and a written submission should be
made to the Publishers.

A CIP catalogue record for this book is available from the
British Library.

Shire Library no. 392   •   ISBN-13: 978 0 74780 764 3

Val Porter has asserted her right under the Copyright,
Designs and Patents Act, 1988, to be identified as the
author of this book.

Designed by Ken Vail Graphic Design, Cambridge, UK
and typeset in Perpetua and Gill Sans.
Printed in China through Worldprint Ltd.

09 10 11 12 13   10 9 8 7 6 5 4 3 2 1

COVER IMAGE
Highland cow and calf.

TITLE PAGE IMAGE
A beef shorthorn bull.

CONTENTS PAGE IMAGE
The 'wild' Chillingham cattle of Northumberland are wary
of humans and are largely self-sufficient.

ACKNOWLEDGEMENTS
The following are gratefully acknowledged for allowing
the reproduction of illustrations (others are from the
author's own sources):

Aberdeen-Angus Cattle Society, page 35 (top); Ayrshire
Cattle Society and Norman Walker, page 39 (middle);
Beef Shorthorn Cattle Society, title page, and page 17;
Belted Galloway Cattle Society, page 37 (top); Blonde
d'Aquitaine Breeders Society, page 47 (top left and right);
British Bazadaise Cattle Society, page 48; British Charolais
Cattle Society, page 45 (both); British Simmental Cattle
Society, page 49 (bottom); Brown Swiss Cattle Society
(UK), page 51; Devon Cattle Society, pages 20, 26 (top
left); Josie Dew, front cover and pages 9, 37 (bottom),
and 55; English Guernsey Cattle Society, page 31 (top);
Galloway Cattle Society, page 36 (both); Eileen Hayes,
contents page and pages 11, 12 (top), 24 (top), 41
(bottom), and 42 (top); Lincoln Red Cattle Society and
Patsy Bradley, page 29; Luing Cattle Society, page 39
(top); Museum of English Rural Life, pages 6, and 33;
Ann Nicholls, page 28 (bottom, both); Anna Oakford,
pages 13 (bottom), and 15 (middle and bottom right);
Jane Paynter, page 42 (top); Salers Cattle Society of the
UK, page 47 (bottom, both); Douglas Scott, page 30
(bottom); Shetland Cattle Herd Book Society, page 41
(top); Shorthorn Society, page 17 (bottom); Stuart Webb,
page 34 (top), 42 (middle and bottom); R. J. Whitcomb,
page 35 (bottom)

CONTENTS

IN THE EARLY DAYS

DOMESTICATION

CATTLE HAVE BEEN DOMESTICATED for at least eight thousand years. Almost all the domesticants originated from the wild aurochs, a beast with magnificent horns and considerable stature found all over Asia and Europe. European aurochs bulls stood up to 6 feet (2 metres) tall at the shoulder and weighed up to a ton (1,000 kg). Julius Caesar described them in *Gallic Wars* as being nearly the size of an elephant. The animals had long horns spreading outwards and then snaking forwards and upwards. Bulls were black with a pale stripe along the back, white muzzle and curly white hair on the poll; cows and calves were reddish brown. The species became extinct in Britain as far back as the Bronze Age (around 2100–750 BC). The last known wild aurochs, a cow, died in Jaktorów Forest, Poland, in 1627. There have been attempts to 'recreate' the aurochs by crossbreeding various primitive breeds of domestic cattle (notably by the Heck brothers, Heinz and Lutz, in the 1920s), but these recreations can only ever be superficially similar to the original.

At first, domesticated cattle (probably originating in the region of Mesopotamia) were long-horned like the ancestral aurochs, but short-horned cattle reached central Europe by 3000 BC and Britain (by boat) by the Bronze Age. They had become very small and were only around 41–2 inches (105 cm) tall in the Iron Age (around 800 BC to 100 AD); they remained small and scrubby for a long while, but they suited the needs and farming methods of the time.

The huge aurochs, wild ancestor of domesticated cattle, was depicted in Palaeolithic European cave paintings – for example at Lascaux in southern France.

Chillingham cattle in the 1860s, as painted by Sir Edwin Landseer.

Eventually, particular types and colours of cattle became closely linked to different parts of Britain. For example, a thousand years ago the heavy draught Saxon oxen that ploughed the claylands of East Anglia and southern England had longish horns and usually red or dun coats, while in 'Celtic' areas remote from Anglo-Saxon influence the cattle were small, short-horned and often black.

In medieval Britain, most cattle remained small and their population was shrinking because grazing land was being ploughed up for grain. Then came the start of a human population explosion in the seventeenth century and, with a growing demand for meat, huge numbers of cattle were reared in remote parts of Britain and sent to fatten in the Midlands. By the late seventeenth century several regional beef types were common: from Yorkshire to Staffordshire they tended to be large with long horns; in Lincolnshire there were tall, long-bodied, short-horned draught animals; and in Gloucestershire and Somerset blood-red cattle were favoured.

By the middle of the eighteenth century the cattle in Scotland, Wales and Cornwall were of various colours, but often black, with or without a white stripe along the back (a pattern known as 'finchback'). From the Pennines westward there were long-horned finchbacks; in Glamorgan there were middle-horned finchbacks, some of them on the Welsh border with white faces. From Devon to Kent the animals tended to be red; along the east coast of England part-red cattle were preferred and in East Anglia in general there was a mixture of red and white cattle and duns.

Painting by Aster Corbould in about 1879 said to show Suffolk or Norfolk, Sussex, Glamorgan, Hereford, Shorthorn, North Wales, Devon, South Wales and Derbyshire Longhorn.

SELECTIVE BREEDING

In 1709 the first Enclosure Act was passed, which would have an important effect on the formation of breeds. The Enclosure Acts increasingly curtailed free-range grazing on common land – an old practice that had incidentally meant free-range breeding. With cattle more confined, breeding could become more selective and the scene was set for the gradual development of true breeds.

A breed can broadly be defined as a group within a domesticated species that, when bred to another of the same group, will *regularly* produce offspring

that are recognisably of that group, both morphologically and in aptitude. It is a matter of consistent replication from one generation to another. It is a mistake to lump together various populations of, say, red and white pied cattle or all colour-pointed cattle as a breed simply because of similar coat pattern, which is a superficial factor. Some say that a breed is a breed if it has a formal breed society, but this again is misleading. For example, there might be a breed society for 'blue' cattle, but blue cattle can readily be produced by crossing a black cow (e.g. Welsh Black) with a white bull (e.g. White Shorthorn). The blues would only be a breed in their own right if two blue cattle mated together produced blue calves, and so on through several generations. Breed societies exist to promote their breed and maintain a herdbook in which 'pedigree' or purebred animals can be registered; they seek to protect the purity of their breed, though in many cases they also attempt to improve it by introducing genes from other breeds (or from a foreign population of what was originally their own breed) that they believe will make a beef breed beefier, a dairy breed milkier and so on.

The magnificent American Brahman, a long-eared zebu breed developed from humped Indian cattle, has been crossed with several British breeds overseas to create new tropical breeds, including the Braford (Brahman/ Hereford), Bravon (Brahman/Devon) and Brangus (Brahman/ Aberdeen-Angus).

Towards the end of the eighteenth century, several factors prompted more rapid development of the breeds. Above all, the industrialisation of Britain created ever higher demands for beef and dairy produce. Breeding began to be more professional, rather than being merely the whim of a farmer to put his favourite bull with his best-looking cow just because he 'felt' it was the right thing to do. Instead, breeding became scientific, with a definite aim in mind, and the era saw the emergence of the early 'greats' among British breeders, such as Robert Bakewell (1725–95), who developed Longhorn cattle and Leicester sheep into the prime breeds of their time.

Soon cattle breeding became fashionable among the aristocracy, who aimed for the biggest and the best but also sometimes imported the exotic. There is a 1770 painting by an unknown artist of 'A Brahma Cow with a Herdsman and Dog' that portrays a gently humped grey-white cow with short horns and eloquent Indian eyes, owned by the Earl of Fitzwilliam. Lord Salisbury had a Bengal bull and at Woburn there were bulls from Gujarat and Ceylon. In the late 1770s the governor of Bengal presented Lord Rockingham with a herd of grey-white humped Indian 'Brahma' cattle for his estate at Wentworth Park; and in the 1860s Henry Tait, Queen Victoria's bailiff at Shaw Farm, Windsor, posed for a painting with a waist-high heavily humped and horned grey-white Brahma bull from Mysore, both bull and man dwarfed by a Shorthorn bull standing behind them. In 1862 the Maharajah of Mysore presented the queen with a couple of Indian 'zebu' (i.e. humped) cows and a bull, portrayed in watercolour by F. W. Keyl in 1871. A set of engravings made by George Garrard between 1799 and 1814 included a 'Fat Galloway Heifer' exhibited at Smithfield in 1804, with his comment 'being a mixture of Indian'. This was at a time when imported Indian zebu cattle were being used in breeding experiments by leading agriculturalists, but the trend didn't last in Britain. It was not just India: an 1880 photograph shows three Zulu cattle at Shaw Farm, presented to the queen by Lord Wolsey. Other 'decorative' breeds for aristocratic and royal parks included the little Irish Dexter and the small, dainty 'Alderney' cows from the Channel Islands.

Meanwhile British cattle had been growing: their average height had increased to about 47–48 inches (about 120 cm) in the seventeenth century and would reach an average of 54 inches (137 cm) in the twentieth. But in the twentieth century Britain had a substantial export trade of cattle to Argentina, where the demand was for smaller, blockier beef animals, and so the prime British breeds became very short in the leg. Later, breeders veered away from the extremely short legs and eventually began to import large Continental beef breeds to add size to their own animals again.

GROUPING THE BREEDS

Breeds can be grouped by role, by coat colour and pattern, by the length or shape of the horns, or by region and relationship. Roles change according to markets and breeders' whims but, in general, an animal bred for beef is muscular and solidly oblong in shape, whereas a dairy cow is daintier and

The belting pattern, here shown in the Sheeted Somerset (painted by William Shiels), an old breed that was already rare by the mid-nineteenth century. There were various herds of 'sheeted' or 'belted' cattle all over Britain.

While the horns of some British breeds are striking, they cannot match the dramatic horns of African breeds such as this Ankole type, bred by the Tutsi tribes of East Africa.

wedge-shaped (deeper and wider at the back end than the front). Many breeds were originally used as draught animals, and the shape here had more weight in the shoulders (for pulling strength) than in the rump. Breeding for beef encouraged development of the rump, and in the case of the Belgian Blue in Europe this was taken to an extreme in what is termed 'double muscling' – a considerable increase in muscle mass that gives the animal exaggerated muscles much like those of a professional bodybuilder. This is also seen frequently in the Piedmont breed of Italy.

As well as the basic coat colours in cattle there are combinations such as brindle (tawny with streaks), roan (coloured hairs mingled with white ones, giving a 'fuzzy' look), pied (clearly defined colour patches on white), finchback (white along the back; a dark line is known as an eel stripe), belted (with a white band encircling a coloured body) and colour-pointed (white with coloured muzzle, ears, lower legs and tail switch – like a Siamese cat). Colour is a badge of identity rather than a guide to qualities.

Horns are no longer useful guides to a breed as most cattle in Britain are now disbudded (deprived of their horns when they are still calves) – a practical measure, but a sad one, as hornless animals look vaguely unbalanced and characterless. Cattle horns are usually grouped as long, middle or short; they also vary in their direction of growth and degree of curvature and there may be horn differences between bulls and cows. Some breeds are naturally polled: they are born unable to grow horns and this trait can be passed to their offspring.

For this book, the breeds are grouped roughly by region, except for those that have been of major historical interest or economic influence on the national herd. There is a section for imported Continentals that have British breed societies, and it is salutary to remember that the trade goes both ways.

In its heyday, Britain bred cattle that spread empire-wide and then worldwide in huge numbers, and several of those breeds had an enormous influence, especially in North and South America, Australasia and Africa. Yet many old British breeds have become increasingly scarce and some have vanished altogether except as a fragment of the gene pool in other breeds. Others have been rescued from the brink of extinction by the Rare Breeds Survival Trust (a charity) or by a handful of breeders who have remained loyal to their breed. It is hoped that this book will open the eyes of the wider public to the broad range and diversity of breeds that we still have, and that we fight to retain. They used to say, 'Britain can breed it.' Britain still can, if only it would.

RARE BREEDS SURVIVAL TRUST PRIORITY LIST OF BREEDS (JANUARY 2009)

Status	Breeds
Critical (fewer than 150 breeding females)	Aberdeen-Angus (original type)
	Chillingham
	Northern Dairy Shorthorn
Endangered (fewer than 250 breeding females)	Whitebred Shorthorn
Vulnerable (fewer than 450 breeding females)	Irish Moiled
	Lincoln Red (original type)
At risk (fewer than 750 breeding females)	Gloucester
	Hereford (original type)
	Shetland
	White Park
Minority breed (fewer than 1,500 breeding females)	British White
Other native breeds	Aberdeen-Angus; Ayrshire; Beef Shorthorn; Belted Galloway; Devon; Galloway; Hereford; Highland; Jersey; Lincoln Red; Longhorn; Luing; Red Poll; Shorthorn; South Devon; Sussex; Welsh Black
Irish origin (but historically part of British agriculture)	Dexter
	Kerry

LONGHORNS AND SHORTHORNS

COLOUR-POINTED CATTLE

THERE ARE NUMEROUS THEORIES about the origins of colour-pointed cattle. Their unusual coat pattern has for many centuries bestowed them with almost mythical qualities in some cultures (including Celtic). The pattern is seen in Scandinavian, Asian and African breeds and the Podolians of Italy and Hungary, but in Britain it could have arisen as a spontaneous mutation. Colour-*pointing*, in which the extremities (such as ears, muzzle, tail switch and feet) are coloured (e.g. black, red) but the rest of the body is basically white, is one extreme of a coat pattern 'cline' in which the opposite extreme is known as 'colour-sided'. A colour-*sided* animal has coloured body sides, head and legs but with a white line along the back, tail and belly (as in Gloucester cattle). Between these extremes are examples where the colouring on the sides is speckled rather than solid and perhaps the face is white. As the pattern edges closer to colour-pointing, the coloured speckling is reduced to a few scattered freckles.

A Vaynol cow.

The feral white **Chillingham** cattle have been living 'wild' in an enclosed 135-hectare park on the Tankerville family's estate in Northumberland for several centuries and are now extremely rare and inbred. The white coat has red or greyish freckles on the face and neck; the muzzle and inside of the ears are red. They are small animals with rather upright horns, often lyre-shaped in the

White Park bull,
symbol of the Rare
Breeds Survival
Trust.

cows. When the artist Landseer wished to paint them in the 1860s, the Earl of Tankerville tried to kill a bull for him to study, but the bull charged, gored the earl's horse, and tossed and pierced a keeper before being shot dead.

The name **White Park** covers assorted herds of colour-pointed horned cattle kept to decorate the landscape (and sometimes to be hunted) in past centuries. The Dynevor, the Chartley and the Duke of Hamilton's Cadzow herds were the basis of today's White Park breed. Most White Park cattle have black points and all have middle-length or long horns; those of the Dynevor type (from near Llandeilo, south-west Wales) grow gracefully sideways, curving forwards and upwards, while those of the Chartley type

A White Park
cow and her calf.

12

Blickling Jenny, a Norfolk White Polled cow in the 1880s. At the time there was a herd of polled colour-pointed white cattle at Somerford Park in Cheshire, said to be exceptionally good milkers. There were also white polled herds in Norfolk, which had been crossed with Shorthorns and Red Polls.

(originally on the estate of the Earl of Ferrers) spread outwards and then downwards. White Parks show potential for beef and have had their own herdbook since 1974 and a breed society since 1983. They have been exported to North America, Australia, Denmark and France.

Set apart from these White Parks is the once feral **Vaynol** herd, which had roamed semi-wild in parkland in Perthshire before being moved to Vaynol Park in Wales in 1872, where the cattle remained until the estate was sold nearly a century later. The herd had a mixed ancestry that seems to have included Highland, Ayrshire and even Indian cattle as well as 'wild' colour-pointed whites (Vaynol horns are similar to the Chillingham's). Cadzow and Dynevor bulls were used in the herd before 1930. The Vaynol is now classified as critically rare: there were only a dozen in the herd when it was rescued by the RBST in the 1980s, but there are now four times as many. The animals tend to be more timid and primitive than other White Park cattle. The herd has been maintained on the Temple Newsam estate, near Leeds, since 1989. Although traditionally any black calves used to be culled, quite a few individuals are now black rather than black-pointed white.

The **British White** shares the White Park's coat pattern, but is polled (i.e. naturally hornless). The British White's points are usually black but sometimes dark brown or red. The greater the degree of freckling and colour spotting on the head, neck and shoulders, the more likely it is that the blood of other breeds can be found in an animal's pedigree. Polled colour-pointed white cattle were brought from Middleton

British White, descended from the 'White Polled' cattle of the nineteenth century. In the late nineteenth century there had been some good milking herds of British White, but it has since become a beef type.

Hall (near Manchester) to Gunton Hall, Norfolk, by Mary Assheton in 1765 when she married; her ancestors had similar cattle at Whalley Abbey, Lancashire, in the sixteenth century. Some of the Gunton cattle had red points, others black. Their descendants in due course became the British White, and by 1918 most had black points. In the late nineteenth century, when its main area was Norfolk and Suffolk, the breed was known as the White Polled (in contrast to the local Red Poll) and was 'swan white' with coloured points and a long, shaggy coat. In Australia it is known as the Australian White and in the USA as, confusingly, the American White Park (despite its lack of horns). It has also been exported to South America and New Zealand.

The Longhorn that Bakewell had improved for meat and milk became a 'fancy' parkland breed, chosen for its long and variable horns, shown here at Calke Abbey, Derbyshire, in an 1880 painting by W. S. Wood.

THE RED AND WHITES

The story of the Longhorn, Shorthorn and black and whites illustrates how the fortunes of breeds ebb and flow.

The photogenic **Longhorn** breed is often seen in parkland. In medieval times the long-horned cattle of northern England and the Midlands were any colour from black to a very pale roan, with or without a white line along the back. Many were dairy cattle, but in the mid-eighteenth century the

The Longhorn remains decorative but is also a good beef animal and suckler cow. The characteristic horns sweep either forward or out and up, or curve down in a semicircle so that the tips turn in towards the cheek in the space-saving 'bonnet' style.

Leicestershire breeder Robert Bakewell (died 1795) was breeding his Longhorns for beef. He also wanted plenty of fat – great bulges of it in the hindquarters – as food energy for active working men and for tallow to light their homes. There was also a good market for the horns themselves as the raw material for various artefacts. Bakewell's methods involved inbreeding, or at least breeding within the herd; his bloodline soon spread and he also exported his animals to Ireland, across the Atlantic to Maryland in 1788, and to Jamaica.

The Longhorn became dominant in Britain, until it was overtaken by the Shorthorn during the nineteenth century. By 1907 there were only about four hundred registered Longhorns, and soon the breed had almost disappeared. In 1973 it was one of the breeds deemed to be rare in a report leading to the formation of the Rare Breeds Survival Trust; by 1996 there were 206 herds and 2,500 breeding animals. No longer classified as a rare breed, the Longhorn produces lean, high-quality beef in quantity, mainly from forage, and the butterfat content of the milk is high. Its coat is now usually red brindle or rich red, with a white line down the back (finchback) and white patch (hough spot) on each thigh. Occasionally its horns grow in different directions; these animals were known as 'waghorns'.

The **Shorthorn** was already being bred in the sixteenth century by the dukes of Northumberland and probably originated from Anglo-Saxon red and Dutch red and white cattle. By the eighteenth century the two best types were the Teeswaters of north-east England, around Darlington, and the Durhams. Two Darlington brothers, Charles and Robert Colling, developed smaller, more compact Teeswaters as milking Shorthorns.

In Yorkshire, Colling stock was used in the 1790s by the Booth family to breed for meat, and then by Thomas Bates for milk. In Aberdeenshire, Amos Cruickshank started to improve his Shorthorns for beef in the 1820s; many were exported to North America from about 1875 and many more went to Argentina.

The Shorthorn herdbook (the world's first) was opened in 1822, and the Shorthorn Society was established in 1875. Forty years on, it was by far the most populous breed in England, and predominated in most of the British colonies and much of North and South America. Shorthorn blood can be found in countless breeds across the world; it has also

An early engraving (1780) of a Shorthorn ox bred at Blackwell, County Durham. It weighed at least 1,500 kg (1½ tons) and its proportions seem more realistic than most nineteenth-century engravings.

Line-up of Beef Shorthorns, showing the typical range of Shorthorn coat colours.

contributed to the creation of new breeds (especially in being crossed with the American Brahman zebu) such as the Santa Gertrudis and Beefmaster in the United States and the Droughtmaster in Australia.

During the twentieth century the black and whites would displace the Shorthorn in the United Kingdom just as the Shorthorn had displaced the Longhorn. By the 1950s there were only 25,000 Shorthorns in their homeland, and this plummeted to a few thousand by 1980.

The Shorthorn had developed into several different types, nearly all of them sharing the basic coat colours of roan, all-red or red and white. In 1958 separate herdbooks were established for beef and dairy types.

The robust **Beef Shorthorn**, mainly from Scotland and based on the Booth/Cruickshank strain, became a good producer of early-maturing 'baby' beef. It left a major mark on cattle in North America, South Africa and Australia; it is still appreciated there and in South America, New Zealand and the former USSR. Although popular worldwide and despite its size having been increased by judicious use of the Maine-Anjou from France, the Beef Shorthorn was declared a rare breed in the United Kingdom in 1987, but is no longer classified as such: in 2009 there were 7,500 Beef Shorthorn breeding females, and their numbers were steadily increasing.

A reliable old favourite, the dual-purpose **Dairy Shorthorn** was based on Bates/Colling strains and it used to dominate Britain's dairy herd. There are now about 7,000 breeding females and it is still a useful and consistent dual-purpose type, producing a good beef calf as well as a plentiful milk supply. The 'Blended Red and White Shorthorn' is a modernised version of the Dairy Shorthorn, developed from 1969 onwards with an increasingly high percentage of red and red-pied cattle (Red Holstein, Red Friesian, Danish Red, Meuse-Rhine-Issel or Simmental) in the 'improved' breed to boost milk yields and dairy character.

The Dairy Shorthorn is still a good dual-purpose cow.

The **Northern Dairy Shorthorn** of the Dales country (also called the Dales Shorthorn) is usually roan and tends to be longer-legged than the Dairy Shorthorn. It was once ubiquitous on the limestone Yorkshire dales; there were 10,000 registered animals in 1944 and more than 5,000 pedigree heifers were registered in 1952. But the dual-purpose Northern is now critically rare, with only about three dozen registered breeding females.

The all-white **Whitebred Shorthorn** (or simply White Shorthorn) is a separate Border counties breed that has been selected from white strains of the Dairy Shorthorn. A century ago it was known as the Cumberland White. Its breed society was formed officially in 1962 and it is now classified by the RBST as endangered. Whitebred bulls are traditionally used on Galloway cows to produce the popular **Blue-Grey** beef suckler cows (i.e. rearing their own calves) of southern Scotland, or on Welsh Black cows to produce the **Blue Albion**, whose coat is blue (mixed black and white hairs) or blue roan, with or without white. Whitebred bulls are also put to Highland cows to produce 'Cross Highlander' suckler cows.

THE BLACK AND WHITES

The **British Friesian**, based on Dutch cattle imported from Jutland in the eighteenth century, was not dissimilar to the Shorthorn in its dual-purpose role, producing both milk and meat but in greater quantities. In the United Kingdom in 1908 there were fewer than forty herds of Friesians, mostly black and white, but dun, dun and white, or black were also acceptable. Within five years there were three hundred breeders and by the late 1940s the black and whites (Friesians, Holsteins or Holstein-Friesians – no one was ever sure what

What's in a black and white name: is it Holstein, Friesian or Holstein-Friesian? This modern cow is the dual-purpose British Friesian.

In the early twentieth century the Shorthorn dominated the national herd but by 1920 the 'Dutch' or 'British Holstein' was the largest of the dairy breeds. Its colour could be dun, dun and white, black or black and white (the latter the most popular). The cow shown here was labelled a 'British Holstein'.

to call them, especially when German names became unpopular after the war) had become Britain's dominant dairy breed. The number of dairy cows in the national herd (including all breeds) dropped sharply from 2.5 million in 1996 to just over 2 million in 2006 and continues to decline, but it remains a fact that the vast majority of those cows are black and whites.

The dual-purpose British Friesian type has been giving way to the dairy **Holstein** type, which originated in the nineteenth century from Dutch Black Pied cattle taken across the Atlantic. In Canada the Holstein was developed specifically as a dairy breed, designed to produce huge volumes of milk. Canadian Holstein blood was first introduced into the British Friesian gene pool in the 1940s. The two types are now so mixed together that it is hard to say what proportion of Britain's black and whites are pure Friesian or pure Holstein, though perhaps a quarter are pure British Holstein. The Holstein is capable of producing higher milk yields than the Friesian, but usually of lower quality, and needing greater inputs to sustain yields.

The British Holstein is taller and more angular than the British Friesian and has a larger frame. It is very much a dairy type rather than dual-purpose.

Still the wheel turns: there is the hint of a gentle return to the old Friesian dual-purpose type to meet changing agricultural circumstances in Britain.

WALES AND ENGLAND

WELSH CATTLE are closely linked with the finchback cattle of the bordering English counties, and these in turn merge into a group of red cattle in southern and eastern England. Most are 'middle-horned', with horn length between those of Longhorns and Shorthorns. The short-horned cattle of the Channel Islands are separate and special.

WELSH CATTLE

The cattle of Wales used to be fattened on Midlands pastures in large numbers. They have a long history and were very mixed in days gone by; colours included white with red ears (mentioned by the Welsh king, Howell the Good, in the tenth century), red or black pied, mouse, dun, red, smoky and blue. By the end of the nineteenth century most Welsh cattle were black, and the two main breeds were the relatively pure Anglesey drovers' 'runts' of North Wales (a mountain type, short in the leg, stocky and heavy) and the larger, rangier and milkier Pembroke (including the Castlemartin) of the south, quite heavily improved with Longhorn, Hereford and other breeds.

Anglesey cattle were small and black, with large dewlaps, flat faces and long horns that characteristically turned upwards. They were graziers'

A Gloucester
herd.

The longish coats of Pembroke cattle (a breed painted here by William Shiels) were ideally chocolaty black, occasionally with the odd white marking. The cows had long yellow black-tipped horns and were good milkers.

animals on the undulating 'ancient island of Mona' (the Welsh name for the Isle of Anglesey), from which they were swum across the Menai Straits in the days before there was a bridge to the mainland. In the mid-seventeenth century about three thousand animals a year took the plunge into the fast current of the deep water, and were sometimes swept downstream for 3 or 4 miles, but usually survived.

Agricultural author William Youatt claimed in the 1830s that 'Great Britain does not afford a more useful animal than the **Pembroke** cow or ox'. The meat was 'beautifully marbled', the animals could 'thrive where

The Castlemartin, a colour-pointed white variety of the Pembroke, was sometimes called the White Welsh. This bull was photographed at Lamphey Court, South Wales, in the late nineteenth century.

The Glamorgan was a favourite of George III: his Glamorgan herd was worked and milked at Windsor. Bulls were black with a white belly and back; cows were black to deep brown or red with the same white finchback markings.

others starve'; they were 'one of the best cottager's cows, while it is equally profitable to the larger farmer'; the black Pembroke ox was a 'speedy and honest worker, fit for the road as well as the plough', and the cow (occasionally dark brown rather than black) was a 'very fair milker'.

The diminutive, heavy **Montgomeryshire** ('full red with smoky points') had dwindled to a mere handful by 1919, and was soon extinct. The handsome **Glamorgan** with its white-marked red coat (white finchback and underline and often a white face) had already become rare before the First

The middle-horned Welsh Black has quite large and hairy ears. The thick soft coat (sleek in hot climates) is jet black, sometimes with the old Pembroke's rusty tinge. In the UK the breed has remained purely British, with no input from foreign breeds or even from overseas Welsh Blacks.

The Belted Welsh is black or red with a broad white belt encircling the body.

The White Welsh is colour-pointed (like the old Castlemartin) and has black, bluish or red points and freckles.

World War. The Glamorgan was an elegant dairy cow, sleek of coat and with a fine 'deer-like' head, and possibly with Norman blood dating back to the twelfth century.

In 1905 the Anglesey and Pembroke were combined as the **Welsh Black** and the types gradually merged. It is now dual-purpose and is mainly seen in suckler herds, rearing calves for beef. There are breed societies in Canada, New Zealand and the USA, all dating from the 1970s.

In theory, colours other than black have been bred out of the Welsh but some farmers liked them and continued to breed belted, white, linebacked and coloured varieties, all of which are now very rare.

The Gloucester's deep mahogany coat is set off by the white of its finchback and underline. The Gloucester of the seventeenth and eighteenth centuries was one of Britain's better dairy breeds for cheese; today it is a dual-purpose cow.

FINCHBACKS

The finchbacked red Glamorgan probably disappeared into the **Gloucester**. In 1796 a Gloucester cow called Blossom was instrumental in Edward Jenner's development of the smallpox vaccine. Closely related to the red Devon and Sussex breeds, the elegant, fine-boned Gloucester has veered close to extinction at intervals ever since the 1830s. After reaching a peak in the eighteenth century, numbers began to decline throughout the nineteenth century. By the early 1890s the Gloucester had been reduced to only one herd, belonging to the Duke of Beaufort at Badminton, and very similar to the Glamorgan. Similarly just one herd existed in 1972, when only seventy 'pure' animals were registered. Since then the numbers have built up slowly but it remains a rare breed, classified as being at risk.

In the 1960s the Hereford was still a short-legged, stocky breed, sturdy and honest. A century earlier it had been a reliable source of oxen for cultivating the land.

The white-faced **Hereford**, which formally became a breed in the mid-eighteenth century, is instantly recognisable anywhere in the world. First exported in about 1817 (to North America), this hugely popular and adaptable beef animal has breed societies almost everywhere. It has contributed to the formation or improvement of dozens of breeds worldwide, including many crosses with zebu cattle in tropical regions. Hereford bulls 'stamp' their white face on all their crossbred offspring.

At first it was a large, stolid draught animal; then it was bred for milk and meat as well as work; and finally it has become essentially a beef animal, but retaining enough of its old milking qualities to be the

perfect suckler cow. The handsome Hereford has a comforting presence that exudes kindness and solid reliability. Its rich, dark red coat is highlighted by the unmissable friendly white face, white chest and feet, white underside and tail switch, and often partial white finchback. The spreading horns are of middle length. The breed society was formed in 1878 and there is also now a Polled Hereford.

Changing production methods after the Second World War, based on science, high inputs and high outputs, led to changes in the original breed into a larger framed, longer legged and later maturing animal that could be fed intensively on cereals, especially in North America, and the older type lost ground to the new. In 1996 British breeders recognised that the original early-maturing, pasture-fed horned Hereford was in danger of disappearing, and they formed a club to save it as the 'Traditional Hereford', defined as an animal whose parentage can be traced in every generation back to the 1878 Hereford herdbook. These now include about 500 females spread over some 45 herds. This traditional type is classified by the RBST as being 'at risk'.

Today the Hereford is well known for having a white face and chest. In 1888 the Aylesbury Dairy Company disposed of a herd of so-called Smoke-faced Hereford cattle, similar to the white-faced Hereford except that the white areas were almost black.

THE REDS

The red of the Hereford is seen unbroken in several old English middle-horned breeds. Just before the First World War the **Devon**, or Ruby Red of North Devon, was the second most populous breed after the Shorthorn in Great Britain (admittedly there were ten times as many Shorthorns as Devons) and was being exported to Australia, Argentina and southern Africa. Although from the wet wilds of Exmoor, with its cold, exposed winters, these cattle have done surprisingly well in hotter parts of the world.

The breed began to spread beyond Devon in the eighteenth century, largely because of Thomas William Coke of Holkham Hall, Norfolk. Holkham Devons were exported to the USA, but they were not the first Devons to cross the Atlantic: the Pilgrim Fathers had taken some with them in the seventeenth century as milch cows and there is still a rare Milking Devon in the USA today. Devons went to southern Africa in 1800 and four years later someone crossed Devons with French and Indian cattle; in the twentieth century more serious breeders crossed Devons with Indian zebu to form new hot-climate breeds in Jamaica, Florida, Hawaii and Brazil.

The Devon's thick mossy coat is a dark, rich 'ruby' red. Its middle-length horns spread out gracefully in the cow. The build is suited to the moors and

Top left: The moorland Devon has yellow-orange skin under its 'ruby red' coat. It is mainly seen in suckler herds and in some places is being 'improved' with French Salers blood.

Top right: The Devon's horns, still proudly displayed here a century ago, are of middle length, but are rarely allowed to grow today. In the late nineteenth century there was a larger and coarser 'Somerset' type of the Devon, with more droopy horns.

Right: In 1797 this team of 86 Sussex oxen dragged the Dyke Road Mill for two miles, moving it from Regency Square to a site behind a hotel in Brighton. The Sussex continued to be a renowned draught breed even in the twentieth century.

uplands: it is agile and fine-legged, and the body is compact and tidy, yet it can weigh more than a Hereford.

Moving eastwards, the reds were broadly similar, but seemed to grow larger and less 'tidy' than the Devon. The very dark chestnut **Sussex** was a famous plough breed on the heavy Wealden claylands of Sussex and Kent and the oxen could still be seen in the early twentieth century pulling massive loads of timber. It was a solid, old draught breed and was improved for beef in the eighteenth and nineteenth centuries.

The Sussex has twice as many sweat glands as other European breeds and adapts well to hot climates. It has found a welcome in North and South America, Australia and southern Africa; in Texas it has been crossed with the humped Brahman to create the Sabre.

The modern role of the Sussex is as a suckler cow, or as a sire to give dairy cows some beefy calves.

The **Red Poll** is a dual-purpose, naturally polled East Anglian breed. It was developed in the early nineteenth century by combining the **Suffolk Dun**, a famously high-yielding but small and thin hornless milker, with the beefy middle-horned **Norfolk Red**. The latter was usually red (sometimes black), with a golden circle about the eye, and was not unlike a smaller-scale Devon; Devons were bred locally by Coke of Holkham and the Earl of Albemarle. The lack of horns in the Suffolk Dun, a genetic trait inherited by the Red Poll, possibly arose from breeding with Galloway bulls from Scotland, or from the Norfolk White Polled (later known as the British White).

Like the Devon, the Sussex has a good set of horns when allowed to grow them.

27

This Norfolk Red cow, Starling, was thirty-five years old when painted by an unknown artist at a time when Norfolk cattle were bred for beef. The head of a Norfolk Red bull became the symbol of Colman's Mustard.

Below:
The Red Poll was initially known as the 'Norfolk and Suffolk' or 'Polled Suffolk' breed. In 1845 it was expected to become extinct in the very near future; fifty years later it was thriving in considerable numbers.

The Red Poll was East Anglia's dominant breed for many years and was exported to the American West, where the lack of horns was a bonus when animals travelled by rail. It was one of Britain's major breeds in the 1940s but rapidly gave way to the higher-yielding Friesians in the early 1960s. It eventually became a rare breed and in 1980 the battle to restore its fortunes

began in earnest, mainly as a suckler cow, though some herds are still milked. Numbers have risen steadily and it is no longer classified as a rare breed. There are milking herds in North and South America, Jamaica, East Africa and Australia; in New Zealand it is more often used in suckler herds. The colour is a deep red, with a red udder, and it has a hairy fringe over the forehead.

The big **Lincoln Red** was originally the Lincolnshire Red Shorthorn, developed in the eighteenth and nineteenth centuries from short-horned Old Lincolnshire cows put to Shorthorn bulls. In the nineteenth century it was red and white or occasionally dun. In 1913 spectators at the Royal Show were impressed by 'the size and scope of the matrons and their magnificent bags' (i.e. udders) among the Lincoln Red cows. From 1939 onwards a polled type was created by using Aberdeen-Angus bulls on Lincoln cows and in 1960 the word 'Shorthorn' was dropped from the breed name; today most Lincolns are polled. Meanwhile two distinct types had evolved, a beef and a dairy, but the emphasis from the 1950s swung to beef, and the last dairy type of the Lincoln Red was registered in 1965. It is now a big, hardy, cherry-red beef animal with a large frame – a characteristic its breeders maintained when most other British beef breeds had become blocky, stocky and short in the leg. In the 1980s the beefiness was improved by crossing with various European breeds, and so in 1998 the RBST began to record the rapidly decreasing population of the traditional 100 per cent native type, free from European blood; it is now classified as vulnerable, though the improved main breed is not a rare one.

Originally seen as a dual-purpose Shorthorn type, in the first decade of the twentieth century the 'Lincolnshire Red Shorthorn' had a high milking reputation and considerable size. Today this big breed is polled and bred for beef.

The 'gentle giant' South Devon was once the biggest breed in Britain and is still a large one.

Down in Devon's South Hams, the cattle have links with Channel Island breeds. The **South Devon** shares qualities with the Guernsey and Jersey that include milk rich in butterfat, and also a type of haemoglobin that is found in Asiatic and some African cattle, but rarely in other European breeds. Yet blood-typing shows a close relationship between the South Devon, the 'Ruby Red' Devon and the Hereford, and an even closer one with Germany's yellow Gelbvieh and the Alpine Swiss Brown. The breed has done well overseas, especially in South Africa (where it is crossed with the Africander and Brahman), East Africa, New Zealand, Australia and North and South America.

The 'Hammers' or 'Big Reds', as these South Devon cattle are sometimes called, are between red and sandy in colour and the skin is yellow-tinged. In the nineteenth century they were known for producing large joints of meat to feed the industrial and mining areas of the Midlands and South Wales. The breed had been triple-purpose, but in the 1920s its breeders leaned towards milk production. This was reversed in the 1950s, and it was classified as a beef breed in 1972, though it is still used as a dairy cow in some countries.

Though the South Devon is now a beef breed, attempts have been made to revert to its traditional use as a dairy cow by breeding back to the earlier type using artificial insemination with 45-year-old semen stored by the RBST.

CHANNEL ISLANDS

The short-horned Channel Island breeds are dairy cattle. They have links with breeds in Brittany and Normandy but have evolved largely in isolation on their islands since 1789, when imports were banned. Described in the mid-nineteenth century under 'foreign' breeds, the name **Alderney** or Normandy included cattle imported from the Channel Islands – for example, the breeds now known as the Jersey and Guernsey. At that time the Alderney was decorative, found only in 'gentlemen's parks and pleasure-grounds', but known for the richness of its milk and the quantity of butter that could be made from it. Colours included light red, yellow, dun and fawn, with or without white patches.

The 'golden' **Guernsey** dairy cow was often seen decorating English and Scottish parkland in the nineteenth century, at a time when its colouring was described as 'orange and lemon with white patches'. It is now a productive dairy breed. Guernseys first came into England in the eighteenth century, and a separate English breed society was established in 1884. There are also breed societies in Australia, North America, Brazil and several African countries.

The Guernsey is pied golden brown and white, with rich orange-gold skin, sandy eyelashes, amber hooves and golden creamy milk. Its conformation is typical of a dairy breed.

A Guernsey bull before the First World War, when the breed was described as 'larger, stronger-boned and coarser' than the Jersey but with 'no mean capacity for beef production'.

31

Right and below:
The doe-eyed
Jersey's coat
colours range from
silver and palest
biscuit through the
browns and greys
of Alpine cattle to
chestnut, mulberry
and almost black.
Sometimes the
colours are broken
by white patches.
Most Jerseys have
noticeable hollows
in front of the
hips.

The dainty, aristocratic **Jersey** is smaller than the Guernsey and unmistakable: the calves look like fawns; the cows have dished faces, large come-hither eyes and a 'mealy' or 'deer' muzzle (a pale halo around the dark-skinned nose, also seen in some West African and Alpine breeds and in the original wild aurochs). The coat colours cover a wide range of fawns, shading to darker and lighter areas on various parts of the body. In ancestry Jerseys probably owe much to 'Celtic' cattle, but there is a hint of the tropics and the East about them, and some say that their ancestors migrated thence through Spain and France. And back again: the elegant little Jersey, imported into England in the eighteenth century, has been exported all over the world. The worldwide population is enormous – even without the huge number of cattle that have been improved with Jersey blood and the new breeds created from crossing native cattle with Jerseys. In Nepal it has become a draught animal on the hills, but in most countries it is valued above all for its butter-rich milk.

SCOTLAND AND
IRELAND

SCOTTISH CATTLE

THE CATTLE OF SCOTLAND evolved from a mixture of Celtic and Dutch types (with a dose of hornless Norse imports and English Longhorns and Shorthorns) that were bred for specific environments and markets. Originally they were a motley lot, varying widely in appearance: some were horned, some were polled; most were black but a few were of every conceivable colour.

In north-east Scotland, this mixed bag was improved by imported bulls from about 1760. Even in the sixteenth century there were some good black 'hummel' (polled) cattle in Aberdeenshire, and gradually recognisable types developed – including the polled, short-legged, dark brown or black

An unsigned and undated painting of a 'Small Black Cow', a description that could have applied to many Celtic cattle in Scotland and Ireland.

Buchan Humlies and Forfar's hairy polled **Angus Doddies**. These polled cattle were notably docile and quiet. Most were black, or with a few white spots; quite a few were yellow (ranging from brindle or dark red to silvery yellow). The Humlies and Doddies were combined in the late eighteenth century and recognised as a polled black breed in 1835, in due course named the **Aberdeen-Angus**.

By the 1850s there were more cattle being bred or grazed in Aberdeenshire than in any other district of Scotland, and the breeders were creating a deep, blocky, rumpy beef type that also managed to look aristocratic. An early improver of the breed was Hugh Watson, of Keillor: his polled black cow 'Old Grannie', born in 1824, produced twenty-five calves before she died of old age in 1859. The underside of one of Watson's cows in the 1830s was only about 8 inches (20 cm) clear of the ground.

Even before 1820 the fine beef cattle of north-east Scotland were being exported to Ireland, France and Australia. In the 1870s they went to Kansas and Ontario – the first of what would become a flood into North America, and also to Argentina and South Africa. The polled black Aberdeen-Angus has now been exported to more than sixty countries and is famed worldwide for the quality of its prime beef and its adaptability in a wide range of climates. There is forty times more of the pure breed in the USA (where it is known as the American Angus) than in the UK today.

Since the 1980s Canadian and US bloodlines have been brought in to reintroduce size to the British breed, along with Australasian and Irish imports.

Typically short-legged Aberdeen-Angus of the 1960s.

Herd of modern Aberdeen-Angus in the United Kingdom. The type is now longer, leaner and taller, especially in America and Australasia, and increasingly so in its home country, where it had become short in the leg in the first half of the twentieth century.

Numbers of Aberdeen-Angus had reached a low point in the UK in the 1960s and are still not great, but the bulls are widely used on other breeds and so their influence remains considerable. The RBST recognises the original traditional Aberdeen-Angus, free from imported bloodlines, as being critically rare. There is also a **Red Angus**, which has contributed to the creation of a composite beef suckler known as the Stabiliser in the USA and recently introduced to the UK as well.

The black Aberdeen-Angus has contributed to the creation of a score of new breeds throughout the world, and is also used to introduce the polling gene into horned breeds. The polled Australian **Murray Grey** beef breed originated in Victoria in 1905 when a grey calf was born to a light-roan Shorthorn cow mated with a black Angus bull. The Murray Grey is dark grey

The Red Angus has been selectively bred for colour and for the cows' maternal qualities in the USA and Australasia for nearly half a century.

Like the Aberdeen-Angus, the Galloway is a polled breed and is usually black. One way of distinguishing between them is that in the Galloway the crown of the head should be wide and rounded, rather than slightly peaked as in the Aberdeen-Angus.

to silver, with a dark skin (which protects it in hot climates), and it has spread to Asia and North America; it has in some places been crossed with zebu cattle. There is a Murray Grey breed society in the UK.

The **Galloway** of south-west Scotland is another naturally polled breed. It was formed towards the end of the sixteenth century and was soon found in large numbers in England for fattening. It is said that, during the reign of George III, an imported white Indian zebu cow was once crossed with a Galloway bull, but the rest of that tale is unknown. By the 1780s the polled Galloway resembled a hornless Highland in assorted colours, though not so shaggy. It was never intensively improved, but it gradually became a suckler beef type. Usually black (often with a brownish tinge), it looks like an earthier Aberdeen-Angus, with an appealing square and heavily fringed face. This hardy hill breed – out in all weathers on rough grazing – produces good beef cheaply and the cows calve easily even when crossed with big Continental bulls.

Dun Galloway cattle. There is also a belted dun.

The **Dun Galloway** is registered in the same herdbook as the black Galloway; red animals are also acceptable. Other colours and patterns used to include brindle, riggit (lineback) and brocket-faced (a large white blaze down

the face). Belted Scottish cattle were recorded in about 1790 and today the dual-purpose **Belted Galloway** not only has a different coat pattern (black or dun body encircled by a broad white belt), but has also retained more of its original milking ability than the black Galloway and is now recognised as a separate breed, fondly known as the Beltie. It

became rare, but has rapidly increased in popularity in recent years and is now classified as a minor rather than a rare breed. Belties are useful suckler cows.

The **White Galloway** has a white coat, colour-pointed with red, black or dun (possibly from feral park cattle long ago). It has been exported and has breed societies in several parts of the world as well as in the United Kingdom (formed in 1877). In the eighteenth century Lord Darlington had a herd of Galloway-type cattle, described as 'finely globed with red and white'; in 1821 William Cobbett said that Lord Carnarvon's herd at Highclere in Hampshire were white hornless 'Galways' with red or black spots – could this be a misspelling of Galloways?

The distinctively horned and shaggy (West) **Highland** cattle are exceptionally rugged and ideally suited to their native uplands in north-west Scotland. These 'Kyloe' cattle used to be swum across the narrow straits between the islands and the mainland. Britain's third smallest breed, Highlands grow larger and faster on better lowland pastures. They have long, widespread 'handlebar' horns, and thick coats of long matted hair in a range of whole colours from cream to black, but usually in shades of tawny dun to reddish brown. With their striking looks, Highland cattle have found new homes in North America, Australia, Russia, Sweden and Germany.

The **Luing** (pronounced 'Ling') was created by the Cadzow brothers on an island off Scotland's west coast by crossing Beef Shorthorn bulls and Highland cows. The new breed was officially recognised in 1965 and has been widely exported.

Unusually for Scotland, the **Ayrshire** is a dairy breed. Originally many cows in Ayrshire were of the Highland or Galloway type but in the mid-seventeenth century the Cunningham

The white belts of Belted Galloway cattle are easily visible from afar on the moors and hills in murky weather. A white-blazed or 'brocket-faced' Galloway used to be quite common in parts of old Wigtownshire.

The Highland, a uniquely hairy breed descended from strong-legged Scottish mountain cattle.

In winter the Highland's long coat should 'cover the brow like a sporran' and it should also have a soft undercoat or 'vest' that is 'so closely set that a bucket of water thrown over the "roof" of an animal would still leave the skin dry'.

region in northern Ayrshire was noted for its Dunlop dairy cattle, and these became known as Ayrshires by the early nineteenth century. From the 1840s they were improved with the help of Dutch, Holderness Shorthorn and other cattle.

The Ayrshire is an attractive cow, with its smart brown-and-white coat, tidy udder and lyre-shaped horns. In the nineteenth century it became popular in the show-ring, where 'pretty' cows, small and delicate, were preferred to practical productive milking cows. Thus, even after the Ayrshire Cattle Society was formed in 1877, there were two types: those bred for show and those bred for milk production, but rarely for both. Eventually the two combined.

By the beginning of the twentieth century Ayrshires were being exported in large numbers to Scandinavia and Canada, and also to the USA, Australia, New Zealand and Japan. The Finnish Ayrshire established itself in huge numbers in the former USSR. In more recent years the Ayrshire has spread to Africa and south-east Asia as well as many European countries. Its numbers remain high in the United Kingdom.

The original **Shetland** was an old island type of crofter's cow. Like other Shetland animals, the cattle were small, and the size was partly due to the environment: in better climates and on better grazing Shetland cattle can weigh anything up to 50 per cent more than island cattle. As in other Scottish breeds, its colour range used to be considerable, including solid red, dun, black, yellow, blue, brindle or grey, or most of these pied or otherwise marked with white. The Shetland is now sometimes misidentified as a Friesian because of its black and white pied coat – a colouring adopted as the breed standard at a time when the Friesian was in fashion.

Top right: The Luing is mainly a 'cow breed' for suckler herds. Its colours are Highland reddish brown, or Shorthorn red and white or roan. Luing cows are sometimes crossed with Simmental bulls to produce white-faced Sim-Luing suckler cows.

Middle right: A hundred years ago the Ayrshire was said to be the finest dairy breed in the world. It has been widely exported, and has also contributed to various foreign breeds.

Bottom: The elegant horns of the Ayrshire are rarely seen today. In the early years of the twentieth century show-ring Ayrshire cows were often mostly white, with only small brown markings on the head and neck; there were also considerable numbers of black and white Ayrshires.

There has been at least a century of crossing with various other British breeds – to such an extent that the pure Shetland was almost lost in the 1950s, but a few animals were rescued from the brink of extinction. It was critically rare, but the breed society was revived in 1982 and numbers began to increase during the 1990s. A group of mainland breeders was established in 2000 and the breed has since done so well that it is now classified as only 'at risk'. It remains a dual-purpose breed, good as a frugal suckler cow and a smallholder's house cow, and it is also used for grazing in conservation areas.

IRISH CATTLE

Archaeological evidence shows that a type similar to the **Kerry** cow has been known in south-west Ireland for some four thousand years. The Kerry has a common ancestry with Iberian cattle, and it has a marked likeness with other remnant 'Celtic' breeds such as the feral Carmargue cattle of southern France, and the fiery little Hérens fighting cows of the Alps.

Kerry cows came to England during the nineteenth century, mostly to decorate rich men's gardens: they are pretty animals, with middle-length horns. The first Kerry herdbook was published in 1890, and a joint breed society with the Dexter was formed in Ireland in 1917; they went their separate ways two years later. Societies were formed in the United Kingdom in 1892 and the USA in 1911. In Ireland the Kerry was almost extinct by

Painting of the 'Zetland' breed by William Shiels (1785–1857), showing that even then black and white was characteristic of Shetland cattle.

Left: The compact, short-legged little Shetland cow has a hint of the fineness of a Jersey and also similarities with old Norwegian cattle. Most Shetlands are black and white but red pied, grey and brindled animals are now being seen again.

Middle: Some Shetland breeders are deliberately selecting red and white animals, as well as black and white, and are also successfully crossbreeding their easy-calving cows with Continental bulls such as the Simmental.

1983, but there are now more than 100 registered herds there, with birth registrations increasing each year, and there are other small herds in the UK and North America. The dainty and elegant black Kerry has a dairy conformation; she is graceful and alert, yet hardy and thrifty – an ideal smallholder's cow on hill farms and with good milk yields.

There is a close relationship between the Kerry and the far more widespread **Dexter**, once known as the Dexter-Kerry. The Dexter was

Below:
The Kerry has a long history as a small Celtic mountain cow in Ireland, very much the family cow for smallholders and cottagers.

Although the Kerry is now black, the colour range used to be wider and included brown, black and brown, black and white and a white-backed black Drimmon variety. The Earl of Clare's Kerry cow painted by William Shiels in the 1830s was red.

probably developed from very small Kerries, and subsequently selected for smallness – to such an extent that some became excessively short in the leg and risked carrying a lethal dwarfing gene. With good marketing the Dexter has risen far beyond its former status as a rare breed and its numbers continue to grow. There are breed societies in Ireland, England, the USA and South Africa (where the demand is particularly keen). The Dexter is a dual-purpose breed: some are milked regularly in dairy herds, others are bred for small joints of good beef; many are housecows, supplying milk for the home. Most Dexters are black, but red and dun are also acceptable.

Dexter cow photographed in the years before the First World War. Her name was 'La Mancha Hard to Find'...

Modern Dexters remain small but in proportion. (Above) Well-bred dairy type in a herd that was regularly milked. (Right and below right) Most Dexters are black, but red and dun or 'golden' are increasingly popular.

Not all Irish cattle are horned; there is archaeological evidence of hornless cattle going back more than three thousand years in Ireland and there are references to 'maol' cattle in traditional stories from the fourth century. The term *maol*, or *maoile*, means 'little mound' or, with reference to cattle, polled. Most of the polled cattle used to be in north-west Ireland and in more recent times they included dairy cattle such as the **Irish Dun** and the nineteenth-century brown-eared **Donegal Red**. They were not really breeds, simply polled cattle, but in 1926 breeders formed a society and opened an **Irish Moiled** herdbook for polled cattle. These did well enough as dairy cattle, but declined rapidly after the Second World War, until by the early 1960s there were only two small herds of Irish Moiled in Northern Ireland. With the encouragement of the RBST a new herdbook was formed in 1983 to save the breed.

The Irish Moiled is a dual-purpose polled breed and usually a rich red with a white dorsal line, white tail and underline, and white udder. As long as the white finchback is present, the rest of the coat may also be yellow dun, roan or plum; white with red ears is acceptable too. Although it has the Longhorn's colouring, it is hornless.

The Irish Moiled remains vulnerable but its numbers are increasing: in 2008 the breed society had about 140 members, with more than forty breeders in England, Scotland and Wales and the rest in Northern Ireland and Ireland.

NEW 'BRITISH' BREEDS

I<small>N RECENT YEARS</small> it has been fashionable to import Continental breeds. All those described here have their own breed societies in Britain, many established in the early 1970s, with another rash of them more recently, and many were introduced to 'beef up' the native cattle.

The majority of these new 'British' breeds are from France, most of them originally draught breeds but now beef. Their claim to Britishness is either that the type has been adapted for British farming and can be distinguished from the foreign original, or, in other cases, simply that a separate breed society has been formed in the UK.

The big wheaten, cream or white **Charolais** from central France was the first to arrive; it was initially imported in the late 1950s when dairy farmers were looking for bulls to make their calves beefier. A British breed society was established in 1962 and the Charolais soon became a leading breed as a terminal sire for crossbred beef calves. The kind-eyed cows are seen in purebred suckler herds, or as mothers of crossbreds. In North America in particular the Charolais is crossed to create new breeds, and has even been crossed with the American bison.

Above and below: British Charolais. In the 1850s France's Charolais was rejected as being half a century behind British breeds of the time; 120 years later it was winning prizes at the Royal Show and boosting Britain's beef production.

Right and below:
The Limousin first
came to Britain
in 1971. There
are now probably
more Limousin
beef animals, pure
or crossbred, in
Britain than any
other beef breed.

The mark of the golden-red **Limousin**, another beef breed from central France, is also seen in many crossbred British suckler herds. The Limousin has middle-length horns and its colour is wheat red, with lighter colouring around the eyes and muzzle and on the lower legs, underside and tail switch. There is a black variety as well; the calves are born light fawn or brown and gradually darken to full black when mature.

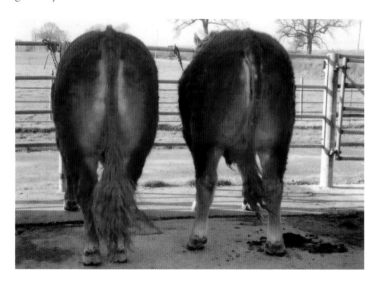

Above: British Blonde bull. Its ancestors included the Garonnais, which possibly reached England in the fourteenth century, and might have been an influence on the South Devon.

Right: Crossbred calves fathered by a Blonde bull, with their Holstein-Friesian mother.

The corn-coloured Blonde d'Aquitaine from south-west France was not recognised as a breed until 1962; it resulted from combining various pale Pyrenean meat and draught breeds. It is called the **British Blonde** in the UK, where it is essentially a beef breed.

The **Maine-Anjou** of north-west France originated in about 1840, when Durham Shorthorns were crossed with local Le Mans cattle. It is dual-purpose, with the emphasis on meat rather than milk, and is red or red and white and sometimes red roan. In 1976 it was approved for use by British breeders to improve muscling in the Beef Shorthorn.

The **Salers** (pronounced 'Salairs'), first imported from France in 1984 to establish a Cumbrian herd of sixty-four animals, is a mountain beef breed with quite long horns from Cantal in the Auvergne. It has spread widely since a British breed society was created in 1986, which now registers about 1,000 animals a year.

Above and right: The Salers is dark conker-red and horned, but there is an increasing proportion of polled black animals within the breed in Britain. In the 1850s the breed in France was improved with Shorthorn, Highland and Devon cattle; in repayment the Devon has recently been boosted by an infusion of Salers.

A newer introduction is the grey **Gascon** of south-west France. The Gascon may have black points ('*à muqueuses noires*') or may not (the '*aréolé*' type, known as the Mirandais in France). It is rare in its homeland. In Scotland it has been used successfully to beef up native Highland cattle.

Another grey French breed is the **Bazadaise** (in France 'Bazadais', pronounced 'Baz-a-day'). Long established as a working animal around the town of Bazas, and with its own French herdbook since 1895, it first came to Britain in 1989 and is bred for beef.

The fawn-coloured **Parthenais** from the Deux-Sèvres region of mid-west France is a double-muscled beef breed, used in the UK to put a bit of meat into calves from Holstein dairy herds.

The reassuringly solid **Meuse-Rhine-Issel** (MRI) is named from the three rivers in its Netherlands area of origin; it is also known as the MRY (the Dutch spelling is Yssel). It originally absorbed several old Dutch breeds, including the Hollander, Zeeland and Drentish. The MRI first came into the UK in the early 1970s.

The good-natured, white-faced **Simmental** originated from western Switzerland and its ancestor was the local Bernese. The pied coat is yellowish-red and white, with white legs and tail. It is used for lowland and upland beef, either in purebred herds or as crossbreds. It is an important breed worldwide and has had its own herdbook in Switzerland since 1806. The Simmental group includes various national Red Pied or Red Spotted ('Fleckvieh' in German) breeds; in addition, many new breeds in Australia,

The coat of the grey Bazadaise is generally darker around the head but the muzzle is pale. The calves are born pale fawn but those from Holstein cows bred to Bazadaise bulls are often black.

The Meuse-Rhine-Issel, a red and white pied milk and meat breed, is built like a muscular British Friesian.

the USA, Africa, China and the former USSR are based on the Simmental. There is also a Black Simmental in the UK.

Another Bernese descendant is the dual-purpose **Montbéliard**, from France's Haute Saône-Doubs region, with a bright russet red and white coat. It came into France with the Mennonites in the 18th century and is also known as the French Dairy Simmental. In its own region it is popular for cheeses and smoked sausages, and in the United Kingdom it is increasingly used in cross-breeding programmes.

The Bernese was also the ancestor of the German **Gelbvieh** (the name means 'yellow cattle' and the colour ranges from cream to reddish), in this

The Simmental's basic coat colour is sometimes described as dun-red or leather-yellow. The trademark white face is passed by the bull to all its crossbred offspring.

Opposite: The Brown Swiss came to Britain from the Alps via North America, and has developed into a distinctive British type in recent years.

case crossed with local Bavarian cattle in the late eighteenth century. It is now more a beef animal than a dual-purpose one, and has contributed to the 'Stabiliser' composite beef 'breed' (along with Red Angus, Simmental and Hereford).

Based on the traditional stocky dual-purpose Swiss Brown of the Alps, the grey-brown American **Brown Swiss** has been bred as a taller and more rangy dairy type since the 1880s. It was introduced into the United Kingdom from Canada in the mid-1970s, later boosted by imports of the old dual-purpose type from Switzerland. The British cow today is less extreme as a dairy type than the American and there are now about 6,000 milking Brown Swiss cows in the UK.

The white, blue roan, black and white or black **British Blue** was originally imported in the early 1980s as the Belgian Blue, a breed that had developed from a mixture of nineteenth-century Durham Shorthorns from Britain and Dutch Black Pied bulls crossed with local Belgian red or red pied cows. By the 1960s it had become an extreme beef type in Belgium, with exaggerated 'double muscling' in the bulging thighs and shoulders. Cows had considerable difficulty in calving and usually had to undergo caesarean sections. A British breed society was formed in 1983, and set about breeding a less extreme type, concentrating on reducing the rate of difficult births. It managed to slash the rate of non-elective caesareans in the British type from 81 per cent in the early 1980s to 17.5 per cent by 2007, and increased the proportion of cows calving easily from 1.5 per cent to 52 per cent in the same period, and it continues to improve on those figures as a priority. An easy-calving 'British type' was recognised in the 1990s, and its name was changed to British Blue in 2007.

Another double-muscled breed brought into the UK in the late 1980s is the Italian white Piedmont or **Piemontese**, with its black muzzle, ears, tail switch, hooves and beautiful black eyes. Italy's Podolian beef breeds (white coats shading to grey over black skin) have also been imported, including the rustic **Romagnola**, the tall and athletic **Chianina** (used in the USA to create the Chiangus and Chiford by crossing with Aberdeen-Angus and Hereford) and the upland **Marchigiana**. These are used as sires for crossbred beef, but numbers remain small in the United Kingdom.

Gradually, some of the Continental breeds described in this chapter are diverging to become British breeds. Perhaps others might soon join them, especially with a changing climate. Maybe the nineteenth-century experiments with Indian humped cattle will be revived in Britain, and perhaps some of the promising African breeds will also find a home here one day. Cattle breeds are constantly evolving, as they always have done, and who knows what other interesting bovines might eventually be proud to call themselves British?

FURTHER READING

Curran, P. L. *Kerry and Dexter Cattle and Other Ancient Irish Breeds: A History.* Royal Dublin Society, 1990.

Hall, Stephen J. G., and Clutton-Brock, Juliet. *Two Hundred Years of British Farm Livestock.* British Museum (Natural History), 1989.

Jones, Professor C. Bryner (ed). *Live Stock of the Farm. Volume I: Cattle.* Gresham Publishing Company, 1920.

Kerry Cattle Society. *Kerry Cattle: A Miscellany.* Kerry Cattle Society of Ireland, 2000.

Moncrieff, Elspeth, with Stephen and Iona Joseph. *Farm Animal Portraits.* Antique Collectors' Club, 1988.

Porter, Valerie. *Cattle: A Handbook to the Breeds of the World.* Christopher Helm/A & C Black, 1991; reprinted Crowood Press, 2007.

Porter, Valerie. *Cows for the Smallholder.* Pelham, 1988.

Porter, Valerie (ed). *Mason's World Dictionary of Livestock Breeds, Types and Varieties, 5th edition.* CABI Publishing, 2002.

Porter, Valerie. *Practical Rare Breeds.* Pelham, 1987.

Porter, Valerie. *The Field Guide to Cattle.* Voyageur Press, 2008.

Stanley, Pat. *Robert Bakewell and the Longhorn Breed of Cattle.* Farming Press, 1995.

Stuart, Lord David. *An Illustrated History of Belted Cattle.* Scottish Academic Press, 1970.

Wallace, Robert. *Farm Live Stock of Great Britain.* Crosby Lockwood, 1893.

Wood-Roberts, John. *Shorthorns in the 20th Century.* Whittet Books, 2005.

Youatt, William. *Cattle: Their Breeds, Management, and Diseases.* Edward Law, London, 1858.

VIDEO
Beef Breeds of Britain. (Narrated by Joe Henson.) Farming Press Videos, 1996.

ORGANISATIONS

Rare Breeds Survival Trust, National Agricultural Centre, Stoneleigh Park, Kenilworth, Warwickshire CV8 2LG.

Telephone: 024 7669 6551.

Website: www.rbst.org

Provides full details of all rare breeds and their breed societies; also details of rare breed farm parks open to the public, and butchers supplying rare breed meat. The quarterly journal for members is *The Ark*.

National Beef Association,

Website: www.nationalbeefassociation.com

For beef breed societies.

Centre for Dairy Information,

Website: ukcows.com/theCDI

For dairy breed societies.

Addresses for cattle breed societies might change frequently where the society secretary is a farmer with an interest in the breed. However, several societies are more permanently based at the National Agricultural Centre, Stoneleigh (Beef Shorthorn, Dairy Shorthorn, British Blonde, British Charolais, British Limousin, British Simmental, Murray Grey) or at Scotsbridge House, Scots Hill, Rickmansworth, Hertfordshire (Guernsey, Holstein, Jersey). The following also have permanent offices: Aberdeen-Angus (Perth), Ayrshire (Ayr), Galloway (Castle Douglas), Hereford (Hereford), Highland (Stirling), Kerry (Killarney), Lincoln Red (Lincoln), Red Poll (Woodbridge, Suffolk), Sussex (Robertsbridge), and Welsh Black (Caernarvon). Many breed societies have their own website for up-to-date information.

The Jersey Creamer Company's Self-Acting Cow-Milker (price 6s 6d), illustrated in the 1890s.

PLACES TO VISIT

Cattle breeds are on display at agricultural shows all over Britain, mostly during the summer. For a list of show dates, consult *Showman's Directory*, Lance Publications, Park House, Park Road, Petersfield, Hampshire GU32 3DL (telephone: 01730 266624. Website: www.showmans-directory.co.uk) or contact the Association of Show and Agricultural Organisations, The Showground, Shepton Mallet, Somerset BA4 6QN (telephone: 01749 822200). Of particular interest is the Rare Breeds Show at the Weald & Downland Museum, Singleton, West Sussex (www.wealddown.co.uk).

RARE BREED FARM PARKS
A selection of centres and breeders with several rare cattle breeds (contact the Rare Breeds Survival Trust for more). Check for opening times, and to confirm breeds before planning a visit.

Cotswold Farm Park, Guiting Power, Cheltenham, Gloucestershire GL54 5UG.
    Telephone: 01451 850307.
    Website: www.cotswoldfarmpark.co.uk
    Belted Galloway, Gloucester, Highland, White Park and others.
Croxteth Home Farm, Croxteth Hall and Country Park, Liverpool L12 0HB.
    Telephone: 0151 233 6910.
    Website: www.croxteth.co.uk
    British White, Dexter, Gloucester, Irish Moiled, Longhorn, Red Poll, Shetland, White Park.
Cruckley Animal Farm, Foston-on-the-Wolds, Driffield, N. Humberside YO25 8BS.
    Telephone: 01262 488337.
    Website: www.cruckley.co.uk
    Gloucester, White Park.
Newham Grange Leisure Farm, Coulby Newham, Middlesbrough TS8 0TE.
    Telephone: 01642 300202.
    Website: www.middlesbrough.gov.uk
    British White, Beef Shorthorn.
Palacerigg Country Park, Cumbernauld, N. Lanarkshire G67 3HU.
    Telephone: 01236 720047.
    Website: www.northlan.gov.uk
    Shetland, White Park.

Sherwood Forest Farm Park, Lamb Pens Farm, Edwinstowe, Mansfield,
Nottinghamshire NG21 9HL.
Telephone: 01623 823558.
Longhorn, Red Poll.

South of England Rare Breeds Centre, Highlands Farm, Woodchurch, Ashfort,
Kent TN26 3RJ.
Telephone: 01233 861493.
Website: www.rarebreeds.org.uk
Beef Shorthorn, British White, Gloucester, Longhorn.

Tatton Home Farm, Tatton Park, Knutsford, Cheshire WA16 6QN.
Telephone: 01625 534431.
Website: www.tattonpark.org.uk
Dexter, Red Poll, Shorthorn.

Temple Newsam Home Farm, Temple Newsam Estate, Leeds LS15 0AD.
Telephone: 0113 264 5535.
Website: www.leeds.gov.uk/templenewsam
Beef Shorthorn, Belted Galloway, British White, Gloucester, Irish
Moiled, Kerry, Red Poll, Shetland, Vaynol, White Park.

Wimpole Home Farm, Wimpole Hall, Arrington, Royston, Hertfordshire
SG8 0BW.
Telephone: 01223 208987.
Website: www.wimpole.org
Gloucester, Irish Moiled, Longhorn, Shetland, White Park.

A Highland
youngster.

INDEX

Page numbers in italics refer to illustrations